OUR PLANET EARTH

Forests

by Karen Latchana Kenney

BLASTOFF! READERS 3

BELLWETHER MEDIA • MINNEAPOLIS, MN

Blastoff! Readers are carefully developed by literacy experts to build reading stamina and move students toward fluency by combining standards-based content with developmentally appropriate text.

Level 1 provides the most support through repetition of high-frequency words, light text, predictable sentence patterns, and strong visual support.

Level 2 offers early readers a bit more challenge through varied sentences, increased text load, and text-supportive special features.

Level 3 advances early-fluent readers toward fluency through increased text load, less reliance on photos, advancing concepts, longer sentences, and more complex special features.

★ Blastoff! Universe

Reading Level

Grade K

Grades 1–3

Grade 4

This edition first published in 2022 by Bellwether Media, Inc.

Library of Congress Cataloging-in-Publication Data

Names: Kenney, Karen Latchana, author.
Title: Forests / by Karen Latchana Kenney.
Description: Minneapolis, MN : Bellwether Media, Inc., 2022. | Series: Blastoff! readers: our planet earth | Includes bibliographical references and index. | Audience: Ages 5-8 | Audience: Grades 2-3 |
Summary: "Simple text and full-color photography introduce beginning readers to forests. Developed by literacy experts for students in kindergarten through third grade"-- Provided by publisher.
Identifiers: LCCN 2021045047 (print) | LCCN 2021045048 (ebook) | ISBN 9781644876046 (library binding) | ISBN 9781648346156 (ebook)
Subjects: LCSH: Forests and forestry--Juvenile literature.
Classification: LCC SD376 .K46 2022 (print) | LCC SD376 (ebook) | DDC 634.9--dc23
LC record available at https://lccn.loc.gov/2021045047
LC ebook record available at https://lccn.loc.gov/2021045048

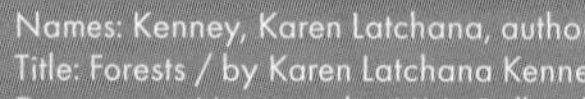

Editor: Kieran Downs Designer: Laura Sowers

Printed in the United States of America, North Mankato, MN.

Table of Contents

What Are Forests?	4
Plants and Animals	12
People and Forests	16
Glossary	22
To Learn More	23
Index	24

What Are Forests?

Forests are large areas where many trees grow. Many other plants grow between the trees.

Forests cover much of Earth's surface. They are home to many different plants and animals.

Boreal forests grow in the far north. Most trees in these forests are **evergreens**.

The temperatures get very cold during long winters. Summers are warmer, but shorter.

Scandinavian and Russian Taiga

Famous For

- One of the largest boreal forests in the world
- Has Norway spruce, Scotch pine, and silver birch trees

Europe

= Scandinavian and Russian Taiga

Type

boreal forest

Size

- 834,745 square miles (2,161,980 square kilometers)

Temperate forests are found around the world. The largest areas are in North America, Europe, and Asia.

Eastern Deciduous Forest

Famous For

- Largest deciduous forest in North America
- Mostly oak, maple, beech, hickory, basswood, and poplar trees

United States

=Eastern Deciduous Forest

Size

- About 926,000 square miles (2.4 million square kilometers)

Type

temperate forest

Some trees in temperate forests are **coniferous**. But most are **deciduous**.

Tropical forests are found around the **equator**. They are warm and **humid**.

Tropical forests have layers. The dark forest floor gets little light. It is blocked by the high **canopy**. The **emergent** layer reaches up to the sky.

Plants and Animals

northern cardinal

Tall maple and oak trees fill temperate forests. Great horned owls and cardinals fly from branch to branch.

Deer and foxes run through shrubs below. Spiders and frogs move across grasses and flowers on the ground.

white-tailed deer

Spruce and pine trees fill boreal forests. Finches eat seeds from their pine cones. Gray wolves hunt moose and caribou.

Kapok trees tower over tropical forests. Toucans nest in the canopy. Coatis move across the forest floor.

People and Forests

People need forests to survive. Plants make the **oxygen** we breathe. Tree roots stop soil from washing away.

People hike and camp in forests. They hunt forest animals. They also use wood from trees to create different things.

building with wood

hiking

Forests are shrinking. People cut them down to make room for homes and farms.

Climate change also hurts forests. **Droughts** make it hard for plants to grow. They also worsen forest fires.

How People Affect Forests

- People cut down forests to make homes and farms

- Climate change causes droughts that make it hard for plants to grow

- Droughts worsen wildfires

Forests help keep the planet healthy. Plant trees to help save forests.

Visitors to forests can keep them clean. They can take all trash out with them. We can work together to keep forests growing!

Glossary

boreal forests—forests that grow in cold regions of Earth just south of the Arctic

canopy—the tropical forest layer beneath the emergent layer; the canopy receives a lot of sunlight and rain.

climate change—a human-caused change in Earth's weather due to warming temperatures

coniferous—related to trees that usually have needles and form cones around their seeds

deciduous—related to trees that lose their leaves in the fall

droughts—long periods of time without rain

emergent—related to the top layer of a tropical forest

equator—the imaginary line that divides Earth into northern and southern halves

evergreens—related to trees that do not lose their leaves in the fall

humid—having a lot of moisture in the air

oxygen—a gas in air that humans and animals need to breathe

temperate forests—forests that grow in mild weather around the world

tropical forests—forests that grow in places that are hot and humid year-round

To Learn More

AT THE LIBRARY

Dorion, Christiane. *Into the Forest.* New York, N.Y.: Bloomsbury Children's Books, 2020.

Kenney, Karen Latchana. *Rain Forests.* Minneapolis, Minn.: Bellwether Media, 2022.

Light, Char. *20 Fun Facts About Forest Habitats.* New York, N.Y.: Gareth Stevens Publishing, 2022.

ON THE WEB

FACTSURFER

Factsurfer.com gives you a safe, fun way to find more information.

1. Go to www.factsurfer.com.
2. Enter "forests" into the search box and click 🔍.
3. Select your book cover to see a list of related content.

Index

activities, 16, 17
animals, 5, 12, 13, 14, 15, 16
Asia, 8
boreal forests, 6, 14
climate change, 19
droughts, 18, 19
Eastern Deciduous Forest, 8
effects, 18, 19
equator, 10
Europe, 8
forest fires, 19
layers, 10, 11, 15
North America, 8
oxygen, 16
people, 16, 18, 19, 21
plants, 4, 5, 6, 9, 12, 13, 14, 15, 16, 19, 20
Scandinavian and Russian Taiga, 7
temperate forests, 8, 9, 12
temperatures, 7, 10
trees, 4, 6, 9, 12, 14, 15, 16, 20
tropical forests, 10, 11, 15

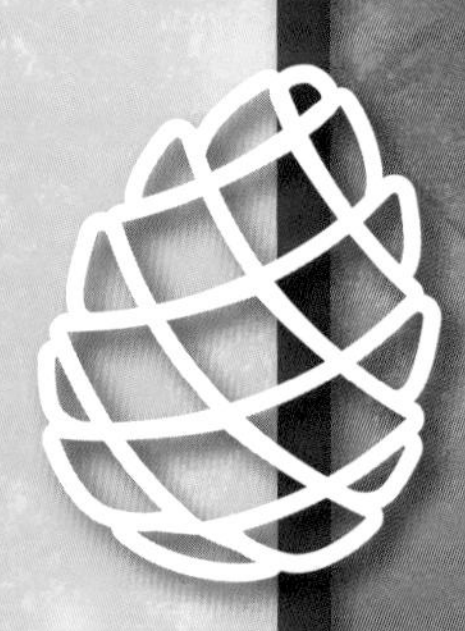

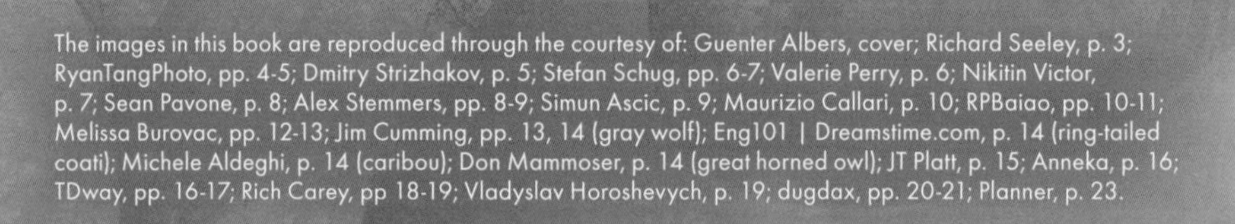

The images in this book are reproduced through the courtesy of: Guenter Albers, cover; Richard Seeley, p. 3; RyanTangPhoto, pp. 4-5; Dmitry Strizhakov, p. 5; Stefan Schug, pp. 6-7; Valerie Perry, p. 6; Nikitin Victor, p. 7; Sean Pavone, p. 8; Alex Stemmers, pp. 8-9; Simun Ascic, p. 9; Maurizio Callari, p. 10; RPBaiao, pp. 10-11; Melissa Burovac, pp. 12-13; Jim Cumming, pp. 13, 14 (gray wolf); Eng101 | Dreamstime.com, p. 14 (ring-tailed coati); Michele Aldeghi, p. 14 (caribou); Don Mammoser, p. 14 (great horned owl); JT Platt, p. 15; Anneka, p. 16; TDway, pp. 16-17; Rich Carey, pp 18-19; Vladyslav Horoshevych, p. 19; dugdax, pp. 20-21; Planner, p. 23.